Michel Ndongo Ndjondjo

Contribution to the study on maintaining food security

Michel Ndongo Ndjondjo

Contribution to the study on maintaining food security

Case of the new Daipn Kinshasa

ScienciaScripts

Imprint

Any brand names and product names mentioned in this book are subject to trademark, brand or patent protection and are trademarks or registered trademarks of their respective holders. The use of brand names, product names, common names, trade names, product descriptions etc. even without a particular marking in this work is in no way to be construed to mean that such names may be regarded as unrestricted in respect of trademark and brand protection legislation and could thus be used by anyone.

Cover image: www.ingimage.com

This book is a translation from the original published under ISBN 978-620-6-72156-7.

Publisher:
Sciencia Scripts
is a trademark of
Dodo Books Indian Ocean Ltd. and OmniScriptum S.R.L publishing group

120 High Road, East Finchley, London, N2 9ED, United Kingdom
Str. Armeneasca 28/1, office 1, Chisinau MD-2012, Republic of Moldova, Europe
Printed at: see last page
ISBN: 978-620-8-18678-4

IN MEMORIAM

To our dearly departed grandparents Cosma YANGO and Marie ASOMBA and our dear cousin John NGONGA, who left early without benefiting from the fruits of their support. May the Lord remember all the good you did us on Earth.

EPIGRAPH

Teach us to number our days, so that we may apply our hearts to wisdom.

Psalms 90:12

DEDICATION

To our beloved parents, who have never ceased to invest in our training and intellectual development.

We dedicate this work.

ACKNOWLEDGEMENTS

We would like to thank the Professor Director of the present work for his availability to read and his encouragement to work.

We would also like to thank all the assistants and other scientific personalities for their support throughout the writing of this work.

We would also like to take this opportunity to express our gratitude to all those with whom we have exchanged ideas on how to improve this work, and for their encouragement in choosing this subject.

To our entire family, we'll never forget all your kindnesses to our modest personality. For this reason, we would like to express our most heartfelt thanks to our parents, who have accompanied us throughout our studies, and to our grandparents, aunts and uncles, cousins and brothers and sisters, for their gratitude throughout this second cycle.

Finally, we'd like to thank all our colleagues and friends from the class and the university as a whole, for the courage they taught us to endure academic realities, and for the exchange of ideas that helped us to master and write this work.

Summary

The DRC has been suffering for decades from growing food insecurity, and this risks becoming a legacy for generations if it is weakly tackled by the Congolese nation through its policy of combating it, despite the fact that the United Nations aims to eliminate hunger in the world by 2030 in order to achieve the sustainable development goals set since 2015.

On the other hand, the country, with its vast, under-exploited territory, is capable of developing all the agricultural strategies necessary to meet the food needs of its people and those of other countries suffering from the same precariousness. It is paradoxical that a country like this should adopt the strategy of importing food instead of producing in its own soil, which is so envied by other countries.

Table of contents

0. GENERAL INTRODUCTION

0.1. Issues

Food insecurity is the real danger troubling the world's population today. This population, despite the enormous difficulties it faces, continues to grow, while at present food resources are proportionally unavailable in every state.

The world's urban population is growing by a million every week, mainly as a result of rural-to-urban migration (ALLEN, 1998). In absolute terms, the total number of undernourished people, or those in chronic food shortage (worldwide), rose from around 804 million in 2016 to almost 821 million in 2017 (FAO et *al.* 2018).

Collective strategies to put an end to world hunger exist and have been proposed by international organizations working in this field, but the assessment of these strategies also depends on the policy of each country that has made this commitment, as its sovereignty matters.

Indeed, food security issues are on the minds of most leaders concerned about the future of their states, and they are initiating plans that in the coming days will help stabilize the balance for every inhabitant in order to tackle the food problem that is disrupting the future of nations.

The DRC has been suffering for decades from growing food insecurity, and this risks becoming a legacy for generations if it is weakly tackled by the Congolese nation through its policy of combating it, despite the fact that the United Nations aims to eliminate hunger in the world by 2030 in order to achieve the sustainable development goals set since 2015.

On the other hand, the country, with all its poorly exploited land, is capable of developing all the necessary agricultural strategies to meet the food needs of its people and those of other countries suffering from the same precariousness. It is paradoxical that a country such as the DRC should adopt the strategy of importing foodstuffs instead of producing in its own soil, so envied by other countries.

For some time now, researchers and analysts have been realizing that, despite the provisions established by the United Nations to completely eradicate hunger in the world, some countries, particularly on the African and Asian continents, will find it difficult to achieve this goal. Other countries, on the other hand, have made efforts to achieve this sustainable development goal (SDG2).

According to FAO (2010), most undernourished people live in developing countries. Two-thirds of them are concentrated in just seven countries (Bangladesh, China, Democratic Republic of Congo, Ethiopia, India, Indonesia and Pakistan), and over 40% of them live in China and India.

Food insecurity is a serious social and public health problem for our cities, provinces and territories, and therefore for the country as a whole, and any solutions to these problems must involve those in power in order to put an end to this disaster, while involving the entire population in the nation's development. In fact, food insecurity implies deprivation in terms of a basic human need: to have access to nutritious food, and in sufficient quantity to remain in good health.

The fact that the DRC is one of the world's most food-insecure countries surprises the rest of the world, since it has millions of hectares of arable land that could solve the food problem, but effective measures to alleviate the problem have long been ruled out, apart from the continual importation of food.

However, during the 70s and 80s, a government policy of large-scale industrial breeding (DAIPN) led to the installation of large complexes near Kinshasa and Lubumbashi.

Today, these large complexes have fallen into a state of lethargy, and the State is now confining itself to its role of supervising breeders and farmers through its specialized services. However, given the ever-increasing needs of urban centers, we are witnessing the gradual emergence of private livestock farming, both on medium-sized farms and in small-scale plots (HUART, 2004).

Household needs are never covered by local production, hence the need to import meat products to meet demand (MPUPU, 2012). New DAIPN's current strategy is designed to solve the country's food problems, although they persist. This persistence is increasingly linked to the lack of secure and sustainable means that need to be put in place to accompany this encouraging vision for the good of the nation.

Thus, we asked ourselves the following questions in order to better conduct our study:

➢ What is the Congolese government's policy on food issues?

➢ Is the Congolese population in general and that of Kinshasa in particular

is she particularly aware of the dietary problems

➢ What is the quality of the food consumed by the inhabitants of Kinshasa?

➢ What are the real causes and consequences of food insecurity?

➢ What strategies does the DRC have in mind to combat food insecurity?

0.2. Assumptions

In relation to the questions posed on the problematic of our study, we believe that :

➢ The DRC has no serious food security policy. Today, the agricultural sector has been abandoned, with no planning, no supervision and, above all, no market.

➢ The agricultural anarchy observed in our country is at the root of undernourishment and food insecurity, the main consequences of which are food imports that pose numerous health problems for the population, and even precipitate the deaths of most Congolese.

> ➤ The New DAIPN has very low production. As a result, it is unable to feed the entire population of Kinshasa. The area farmed by the New DAIPN is insufficient, with limited and mediocre financial, material and technical resources.

0.3. Work objectives

0.3.1. General purpose

Our work focuses on sustainable development through local strategies to combat food insecurity.

0.3.2. Specific objectives

As specific objectives, we set :

- New DAIPN's productivity ;
- Identify the flagship crops of the New DAIPN ;
- Identify sellers and consumers of new DAIPN products
- Propose sustainable strategies for improving food needs.

0.4. Choice and interest of subject

The choice of this topic is to highlight local food security strategies.

The interest of this work is to popularize the New DAIPN as a case study.

0.5. Work delimitation

Food security being so vast and complex, throughout this work we have limited ourselves to what the New DAIPN produces for the food needs of the Congolese people in general and Kinshasa in particular.

In terms of time, our study is carried out from November 2019 to February 2020.

Spatially, our surveys and research focused on the commune of N'SELE, and more precisely on the Nouveau DAIPN site.

0.6. Subdivision of work

Apart from the general introduction and conclusion, this study comprises three chapters:

- ➢ The first deals with the general aspects of food safety;
- ➢ The second is devoted to the study environment, materials and methods;
- ➢ The third and final section presents the results and discussion.

CHAPTER 1: GENERAL INFORMATION ON FOOD SAFETY

1.1. A BRIEF OVERVIEW OF THE FOOD SITUATION IN RDC

1.1.1. FOOD SITUATION

For decades, the DRC has been on the list of countries where food insecurity is rife, according to rankings drawn up by international organizations that have dealt with this problem. As a result, plans and strategies have been drawn up to help this population overcome the obstacle of food insecurity, but alas, this response to need is not so sustainable, given the recurrence of the scourge and the vulnerability caused by food insecurity.

In the Democratic Republic of Congo (DRC), agriculture is the most important sector in terms of occupation and is the most promising basis for achieving food security as well as general economic development (BUGEME and ULIMWENGU, 2019).

The land of the Democratic Republic of Congo represents 227 million hectares, or 97% of the national territory. There are six main soil types: andosols (0.5% of all soils), vertisols (1%), hydromorphic soils (5%), nitosols (ferrisols) (14%), ferralsols (53.5%) and arenoferrals (26%). Truly fertile soils (andosols) occupy a limited surface area. Nevertheless, 80 million hectares are considered relatively suitable for agriculture, of which only 10 million are devoted to crops and pasture (MALELE, 2003).

Despite the country's enormous agricultural potential, the majority of the DRC's population remains largely exposed to food insecurity, malnutrition and hunger. The DRC is one of the few African countries with an enormous potential for the development of sustainable agriculture (in millions of hectares of potential arable land), a diversity of climates, an important hydrographic network, an enormous halieutic potential and a significant livestock potential. Yet the DRC is classified as a Low Income Food Deficit Country (LIFDC). In terms of the

Human Development Index, the UNDP ranked the country 187th out of 187 countries listed in 2011 (WFP, 2014).

However, most food security strategies in the DRC are still aimed at resolving problems in the short term, yet response plans and strategies must address both the short and long term, while ensuring food sustainability. This is where we need to remember that any strategy adopted must always consider future aspects, taking into account likely future impacts, to make the population affected by the scourge less vulnerable in the time to come.

It should be noted that, generally speaking, and in the current policy aimed at reducing and eliminating hunger, having assets alone does not guarantee success, as these assets will need to be transformed to achieve normal success. It is therefore essential to remember that a plan to ensure food security in a given environment cannot dissociate the contribution of the technological revolution to finally produce in sufficient quality and quantity, as well as surplus.

I.1.2. CATEGORIZATION

The policy of the Democratic Republic of Congo has always envisaged the complete elimination of hunger throughout the country, in order to guarantee the well-being and/or food security of its entire population, who have been hoping for this for decades. As a result, despite being endowed with all the resources needed to solve its problems, the DRC continues to experience economic crises, and has always been one of the world's underdeveloped countries, with major humanitarian emergencies.

Yet it's no paradox that a country with an arable land potential estimated at millions of hectares, and with a number of assets for its development, should be experiencing food insecurity for several decades, and is, according to the FAO, one of the seven countries where most of the world's undernourished people live; and these countries account for two-thirds of the world's undernourished people: Bangladesh, China, the **Democratic Republic of Congo**, Ethiopia, India, Indonesia and Pakistan; this is because its policies are geared to short-term responses, taking account of its current economy, and also because it lacks sustainable management of the food sector, which needs to consider long-term solutions.

In a 2010 publication, the FAO also pointed out that in the Democratic Republic of Congo, assessments carried out by aid and development agencies are too often focused on identifying immediate needs, and that the capacities and possibilities of local organizations to intervene in program planning and implementation are frequently ignored.

The food crisis situation in DR Congo has been evolving while stagnating for decades. While many claim that the humanitarian situation in the DRC is 20 years old, the determining factors and consequences of the humanitarian crisis have been present in the country since its independence in 1960 (MUTEBA and NKULU, 2019).

1.1.3. CAUSES

Food insecurity has various causes, but in this work we describe the country's current food situation on three main causes, namely:

1.1.3.1. Political cause

As the main actor in development, the state plays an important role as a public authority in development, even though it shares responsibility in this area with other actors such as private companies and businesses (MUKOKA, 2014).

The policy guides the nation's interest in the choice made for the envisaged sector, presenting all the advantages and opportunities necessary in the days to come for the well-being of the population. It's important to note that anything that is to be carried out must go through a political decision that will ensure the success of all the planned objectives, as well as the promotion and appropriation of the activity concerned by the target population.

With this in mind, the political decision is described as the driving force behind any decision that the leaders of our beloved and beautiful country must initiate to solve current and future food-related problems.

1.1.3.2. Economic cause and other movements

A country's economy indirectly reflects the living and social conditions of its population as a whole; while the country experiences repeated serious economic crises, ambitious development projects are increasingly discouraged and neglected, requiring huge start-up costs and all the necessary follow-up until they become profitable, while the country's situations demand immediate responses.

In areas of the country where insecurity reigns, the situation makes it impossible to envisage an agricultural project to solve a food problem, as uncertainty takes over and mature crops run the risk of being looted or destroyed by the local population, who live in difficult food conditions and have no financial means of their own, or by adversaries.

I.1.2.3. Cause due to research

Research is the first golden rule for solving any problem in today's world. It consists in identifying, analyzing and proposing possible solutions, so that decision-makers can support the resolution of the issues at hand. We must remember, then, that research influences policy to adopt what researchers discover will contribute to the well-being of the population.

A given problem is rarely due to a single source or cause: that's why we need to analyze its social, economic and even political ramifications and causes; because a problem needs to be analyzed in its global context and at different levels (macro, micro, intermediate).

In this categorization of causes, it should be pointed out that the lack of up-to-date guidance from researchers on food policy and agricultural production does not contribute to the country's development or its improvement in food security or the elimination of hunger.

However, while it is undoubtedly necessary to adopt rational economic policies if food security is to be achieved, these are not easy to implement in the absence of real political consensus. In the final analysis, the ultimate responsibility for food security lies with governments, in conjunction with local authorities and in collaboration with the groups and individuals that make up society (FAO, 1996).

I.1.3. CONSEQUENCES

Food problems remain real catalysts for certain others, notably the various diseases that can be encountered on a daily basis in the areas most affected by food problems. If the right measures are not taken to improve dietary conditions while the threat lurks, those most affected are ready to develop a new stage in their state of health until they reach that of malnutrition.

Research has shown that hunger leaves an indelible mark on physical and mental health, particularly in children, which can manifest itself in the form of depression and asthma in adolescence and early adulthood. The problems caused by poor nutrition directly affect all other aspects of life, including the social, environmental and economic spheres.

As pointed out by RIVERA (1991), a population's diet, health and socio-economic stability depend on food, and therefore on an agricultural policy of food production from a nutritional point of view, or better still, a policy of food security.

It's easy to see, then, that in areas (such as the DRC) where there is no agricultural policy (or food security policy), a high proportion of people live in poverty and suffer from food-related problems such as food insecurity, malnutrition and hunger.

I.2. FOOD INSECURITY

I.2.1. DEFINITION AND CHARACTERISTICS

Food insecurity exists when all people, at all times, do not have physical and economic access to sufficient, safe and nutritious food to lead a healthy and active life (FAO, 1996).

It is most noticeable when people are undernourished due to the physical lack of availability of foodstuffs, or economic or social non-access to foodstuffs and/or inappropriate use of food.

It results from prolonged periods of poverty, lack of assets, and inadequate access to productive or financial resources, but can be overcome through typical long-term development measures, which are also used to address poverty issues, such as education or access to productive resources.

Food insecurity can not only cause lasting damage to future generations and to the environment, but can also harm an individual's physical health through malnutrition. It is important to be aware that food insecurity can lead to this serious and potentially fatal condition in the longer term. That said, malnutrition is not always caused by food insecurity. It can result from a multiplicity of other causes, including disease, an unhealthy environment, the consumption of undrinkable water and many others.

I.2.2. TYPOLOGY

There are two types of food insecurity:

- Chronic food insecurity and ;
- Transient food insecurity.

1.2.2.1. Chronic food insecurity (CFI)

According to the FAO World Food Summit held in Rome in 1996, chronic food insecurity implies a permanent deficiency in the diet due to the long-term inability to obtain food.

Food insecurity is said to be chronic when the measures taken to combat food insecurity in a given environment are not so effective as to halt the problem, and insecurity persists. It is therefore caused by the ineffectiveness of the measures established to alleviate the problem.

1.2.2.2. Transient food insecurity (TFI)

As for this type, the World Food Summit points out that it illustrates the temporary inability of a household to access food, due to obstacles linked to food prices, food production or household income.

This type of food insecurity occurs when effective strategies for maintaining food security are absent for a period of time. Transient food insecurity can only be observed after climatic shocks or other movements in an environment where food security already exists.

I.2.3. VULNERABILITY

Vulnerability refers to a group of people who are able to maintain an acceptable level of food security in the present, but who may be at risk of food insecurity in the future.

According to the International Federation of Red Cross and Red Crescent Societies (2005), vulnerability measures the degree of risk to which members of a family or community are exposed when faced with situations that threaten their lives and livelihoods.

A household's vulnerability is determined by its ability to overcome problems and hardships such as drought, flooding, unfavorable government policies, conflict or HIV/AIDS. The severity and duration of the crisis, as well as its timing, are important factors.

She stressed that vulnerability is not synonymous with poverty, although poverty is often an aggravating factor in vulnerability to crises. In other words, crises have more serious consequences when they occur in a context characterized by widespread structural poverty.

ACF-IN (2008), states that, in general, the level of vulnerability of a household and/or individual is determined by the risk of failure of coping strategies. Basic needs are not covered due to a lack of adequate resources (capital, food stocks) and strategies or mechanisms available to the household to cope with a situation or crisis.

Food vulnerability therefore refers specifically to all the factors that place people at risk of food insecurity. The degree of vulnerability of an individual, household or group of people is determined by their exposure to risk factors and their ability to cope with and survive crisis situations (FAO, 1996).

I.3. FOOD SAFETY

I.3.1. DEFINITION

The most widespread and official definition of food security is that proposed by the Food and Agriculture Organization of the United Nations (FAO) at the World Food Summit held in Rome in 1996.

Food security exists "when all people, at all times, have physical and economic access to sufficient, safe and nutritious food for an active and healthy life".

I.3.2. FEATURES

Food safety is characterized by the following four dimensions: Availability, Accessibility, Utilization and Stability.

1.3.2.1. Food availability

Food availability at national, regional and/or local level means that food is physically available because it has been produced, processed, imported or transported. For example, food is available because it can be found in markets, because it is produced on farms or gardens, or because it comes from food aid. It is food that is visible in the environment and corresponds to the food availability carried on the "supply side" of food security, which is determined by the level of food production, provisioning levels and net trade.

Food availability is represented by food supply, which depends, among other things, on the relative prices of inputs and production, as well as on the technologies used for production (FAO, 1996).

1.3.2.2. Food accessibility

Food accessibility is the way in which people can obtain available food. This involves the idea that good food supplies at national or international level do not in themselves guarantee household food security.

Normally, food is accessible through a combination of home production, stocks, purchases, barter, gifts, loans or food aid. Food accessibility is guaranteed when communities and households, Including all the individuals within them, have adequate resources (money, for example) to procure the food they need for a balanced diet. It depends on household income, the distribution of that income within the family, and the price of foodstuffs. It also depends on the social, institutional and commercial rights and prerogatives of individuals, including the public distribution of resources and social protection and welfare systems.

Access to food is influenced by demand, which is a function of a number of variables: the price of the desired foodstuff, the price of complementary and substitute products, income, demographic variables and tastes and preferences (FAO, 1996).

1.3.2.3. Food use

Food utilization is the way people use food, and depends on food quality, storage and preparation, basic nutritional principles and the state of health of the individuals consuming it. Utilization is about how the body optimizes the various nutrients present in food. Good care and feeding practices, food preparation, diet diversity, and food distribution within the household result in an adequate supply of energy and nutrients. This, together with good biological utilization of the food consumed, determines the nutritional status of individuals. Utilization emphasizes the importance of non-food inputs in food security, including sanitation, health care and drinking water consumption in achieving a state of nutritional well-being where all physiological needs are met.

Some illnesses do not allow optimal absorption of food, and the growing disease requires increased consumption of certain foods.

Food utilization is often reduced by endemic diseases, poor hygiene conditions, lack of knowledge of basic nutritional principles, or by traditions restricting access to certain foods on the basis of age or gender. The change in total demand for market services depends mainly on consumer income: the higher the disposable income, the more the consumer can afford market services (GOOSSENS, 1997).

I.3.2.4. *Food stability*

To be food secure, every individual must have access to adequate food at all times, without the risk of losing access to food due to sudden shocks (climatic crises) or seasonal events. This aspect refers to availability, accessibility and utilization. But recourse to external food aid is an easy option for many governments, who see it as an easy way to feed urban populations without having to invest in increasing national food production (Marc, 2004).

I.3.3. FOOD SAFETY ASSESSMENT

According to ACF-IN (2008), a food security assessment represents a given situation at a given time. It is therefore necessary to place the current situation

In time, to date it, in order to better apprehend the risks and levels of vulnerability.

The first step in this assessment is to understand how the community functions and the characteristics of the different household types. Factors such as the general household economy, agricultural systems, health systems, exchange systems, mutual aid systems, household food economy, intra-family ties, and coping mechanisms for particular events need to be examined. Next, the capital available to households and their members, their respective activities and the direct impact of the crisis on the household are the factors studied in order to determine the means (capacities) available to the family to reduce the adverse effects of events.

I.3.4. SOCIO-ENVIRONMENTAL ASPECTS OF FOOD SAFETY

All actions linked to human life are in one way or another closely linked to food, which is recognized as the provider of energy (the power to do work). This being the case, there is a great deal of importance to be attached to this area in order to be able to improve it and envisage the socio-economic development of the nation as well as its well-being in order to achieve sustainable development.

The problem of hunger is a recurring issue in countries where the pace of economic growth and the increase in food production lags well behind the rise in the number of mouths to feed (Marc, 2004). Food security remains one of the means by which a nation like the DRC can take the first step towards its development, by improving the living conditions of its population and involving all its citizens in making this process a reality.

In his address to the 1996 World Food Summit, FAO President Jacques Diouf stated: "Only a society whose daily bread is assured can work effectively for development, as well as for justice, peace, education and other fundamental rights" (FAO, 1996).

The fight against climate change is a global commitment, given that the quality of the environment matters to everyone, and the harmful effects spare no one, no activity whatsoever, and no agricultural practice.

No society in today's world can succeed in combating climate change unless the food problem of its population is solved, as this population will be poised to affect and exploit forests and other resources irrationally, yet the world's leaders have pledged to protect and conserve forests sustainably in order to cope with the effects of climate change and prevent the risk. Finally, it should be noted that food security does not only address food-related issues, but also offers many benefits in other areas of life that interact with it.

1.3.4.1. Social

All activities linked to maintaining food security are primarily aimed at improving the living conditions of the population concerned. Food security is the best way to combat the poverty and unemployment affecting the country's populations.

All the activities involved in this activity generate various opportunities for the intervention of human resources in order to carry them out in the shortest possible time. Taking into account the seriousness of the activities to be carried out to feed a large number of people in urban areas, or the country in general, the workforce is not negligible, and this is what will contribute to poverty reduction.

1.3.4.2. Ecology

Despite the negative impacts that certain activities can have, it is essential to note that the majority of food safety activities are beneficial to the environment.

In fact, it contributes to the conservation of biodiversity, the fight against atmospheric pollution and heat islands. Biodiversity conservation is an important element, as it forms the basis of food security. Biodiversity loss can disrupt ecosystems and make the production of necessary services unstable (CAAAQ, 2007). The primary cause of heat islands in cities is the reduction in vegetation cover associated with densification and urban development. As a result, summer temperatures become hotter and more humid. These heat islands have harmful impacts, such as deteriorating air quality both inside and outside buildings, increasing energy demand for air conditioning and drinking water for cooling, and accentuating human health problems (INSPQ, 2009) cited by GAUDREAULT (2011).

I.3.4.3. The economy

Involving a large number of people in an activity such as employment is so beneficial to the development of the communities concerned, because everyone contributes what they need to. This gain is not necessarily a direct economic one. However, it does promote the integration of individuals into the job market and participation in the global economy (GAUDREAULT, 2011).

I.4. SUSTAINABLE DEVELOPMENT CONCEPT

I.4.1. DEFINITION AND ORIGINS OF THE TERM

Sustainable development is development that meets the needs of the present without compromising the ability of future generations to meet their own needs (WCED, 1987).

This has remained the most widespread and official definition of sustainable development since the publication of the report of the World Commission on Environment and Development (WCED), entitled "Our Common Future" (also known as the Brundtland Report), in 1987.

In 1984, the UN General Assembly mandated a commission of experts, known as the World Commission on Environment and Development (WCED), to propose guidelines for a global development project capable of protecting the environment and fulfilling the other missions included in the development objective.

Chaired by Gro Harlem Brundtland, the WCED brings together some twenty academics and politicians of different nationalities. Its vice-president is Mansour Khalid, former foreign minister of Sudan, and its secretary general is Jim McNeill, former environment director at the OECD. Maurice Strong, a Canadian like McNeill, former secretary of the Stockholm Conference (1972) and later of the Rio Conference (1992), is also a member of the WCED.

It was in this report that the term *"sustainable development"* made its first appearance, but was first translated in Quebec in 1988 as *développement soutenable, and* a few years later remained as it is today, i.e. *développement durable* in French.

I.4.2. CURRENT CONCEPT AND BENEFITS

Sustainable development assumes that all essential human needs are met by everyone, everywhere, in the long term. It is therefore a means by which populations find solutions to their current (and long-term) problems (such as famine, housing, education, employment, etc.), while looking ahead to those of the future.

Indeed, sustainable development also means developing mindsets, behaviors, uses and practices that are mindful of preserving resources and balances, as well as preparing for the lives of future generations.

The main aim of development is to satisfy human needs and aspirations. At present, the basic needs of many people in developing countries are not being met: the need for food, shelter, clothing and work. What's more, beyond these essential needs, these people legitimately aspire to an improvement in the quality of their lives. A world where poverty and injustice are endemic will always be prone to ecological and other crises. Sustainable development means that everyone's basic needs are met, including their aspirations for a better life (WCED, 1987).

Sustainable development is a concept currently used in many of the world's languages, in the media and even in politics. This highlights its great interest and importance in all social classes and strata around the world (NDONGO, 2017).

According to the WCED, meeting basic needs means not only ensuring economic growth in countries where the majority of people live in poverty, but also ensuring that the poorest people can benefit from their fair share of the resources that enable this growth. The existence of political systems guaranteeing popular participation in decision-making, and more effective democracy in international decision-making, would enable this justice to emerge.

If sustainable development is to be achieved worldwide, (according to the WCED report) the affluent must adopt a lifestyle that respects the planet's ecological limits. This applies to energy consumption, for example. What's more, excessive population growth can increase pressure on resources and slow down improvements in living standards; sustainable development is therefore only possible if demographics and growth evolve in harmony with the productive potential of the ecosystem.

However, sustainable development is based on the following three pillars, which we will explain in detail: environmental, social and economic.

1.4.2.1. The environmental dimension

Preserve, improve and enhance the environment and natural resources over the long term, by maintaining the major ecological balances, reducing risks and preventing environmental impacts. Sustainable development is based on the fundamental idea of making human beings aware that nature's resources are not inexhaustible and that the population will continue to grow. By respecting the environment, pollution is reduced and the planet is preserved.

1.4.2.2. The social dimension

In its social dimension, sustainable development aims to satisfy human needs and achieve social equity, by encouraging the participation of all social groups in issues of health, housing, consumption, education, employment, culture, etc.

Through the social dimension, many social problems are resolved, notably unemployment, respect for human beings and the preservation of their rights; it is also the capacity of our society to ensure the well-being of all citizens. This well-being is the ability of each individual to meet essential needs: food, security, health, housing, education, equal access to work, human rights, culture and heritage, etc.

I.4.2.3. The economic dimension

To develop economic growth and efficiency, through sustainable production and consumption patterns, i.e. to promote optimal management of human, natural and financial resources, in order to meet the needs of human communities. The challenges of a sustainable economy are often linked to the other two pillars of sustainable development.

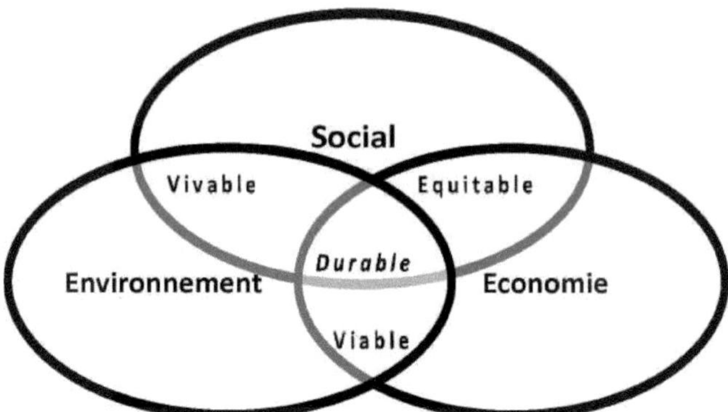

*Figure 1: Sustainable development **diagram***

I.4.3. SUSTAINABLE DEVELOPMENT OBJECTIVES

According to my UNICEF, the Sustainable Development Goals (SDGs) are based on the Millennium Development Goals (MDGs), 8 goals aimed at fighting poverty, launched in 2000, and which the world committed to achieving by 2015.

The eight Millennium Development Goals are as follows: Goal 1: Eradicate extreme poverty and hunger

Goal 2: Achieve universal primary education

Objective 3: Promote gender equality and empower women

Goal 4: Reduce infant mortality

Goal 5: Improve maternal health

Goal 6: Combat HIV/AIDS, malaria and other diseases

Objective 7: Ensure a sustainable environment

Goal 8: Develop a global partnership for development

Enormous progress has been made towards achieving these goals. However, despite these efforts, poverty still affects too many people in the world.

This post-2015 development agenda (SDGs) is much more ambitious and detailed than the previous one (MDGs). The 17 goals to be achieved in 15 years (by 2030) are :

1. ***Eradicating poverty***: in all its forms and everywhere in the world.

2. ***Fighting hunger***: eliminating hunger and famine, ensuring food security, improving nutrition and promoting sustainable agriculture.

3. ***Access to health***: giving people the means to lead a healthy life and helping to ensure the well-being of all at all ages.

4. ***Access to quality education***: ensuring access to education for all and promoting equitable, quality learning opportunities throughout life.

5. ***Gender equality***: achieving gender equality by empowering women and girls.

6. ***Access to safe water and sanitation***: guarantee universal access to water and sanitation, and manage water resources sustainably.

7. ***Renewable energies***: guaranteeing access for all to reliable, sustainable and renewable energy services at an affordable cost.

8. ***Access to decent jobs***: promoting sustained, shared and sustainable economic growth, full and productive employment and decent work for all.

9. ***Innovation and infrastructure***: support small businesses so that they can grow, encourage the development of companies that respect the environment and make healthy products (that don't harm our planet or people), and provide access for all to new technologies.

10. ***Reducing inequalities***: reducing inequalities between and within countries.

11. ***Sustainable cities and communities***: creating cities, housing and transport that are open to all, safe, resilient and sustainable.

12. ***Responsible consumption***: introduce sustainable consumption and

production patterns: avoid waste, reduce waste and consumer goods (books, clothes...) by reducing, reusing and recycling.

13. *Combating climate change*: taking urgent action to combat climate change and its consequences.

14. *Protecting aquatic flora and fauna*: conserving and sustainably exploiting oceans, seas and marine resources.

15. *Protecting terrestrial flora and fauna*: preserving and restoring terrestrial ecosystems, ensuring that they are used sustainably, managing forests sustainably, combating deforestation and desertification, halting and reversing the process of land degradation and halting the loss of biodiversity.

16. *Justice and peace*: promoting peace, ensuring access to justice for all, and building effective, accountable and open institutions at all levels.

17. *Partnerships for global goals*: revitalize the global partnership for sustainable development and strengthen the means of this partnership.

I.4.4. SUSTAINABLE DEVELOPMENT IN THE DRC

The success of the Sustainable Development Goals appears blurred from the outset for the DRC, as the country has by no means made enough effort even to achieve any of the objectives with a view to having a vision in line with that set by the United Nations for achieving sustainable development. But it's high time we had the will, especially the political will, to promote and make an effort to achieve some of these objectives.

As a signatory to the Millennium Development Declaration aimed at "eradicating poverty as the best means of ensuring sustainable development for all", the DRC implemented the MDGs over the period 2000-2015. On the strength of its commitments, and despite a worrying security situation, the country has undertaken several reforms in key areas within the framework of the three

DSCRPs (Growth and Poverty Reduction Strategy Papers) supported by the PAGs (Government Action Programs). Implementation of the MDGs has been monitored through the capitalization of four progress reports (2005, 2010 and 2012, 2015) as well as four MDG Acceleration Frameworks (CAO) in the sectors of food security, education, infant and maternal mortality, and the fight against HIV/AIDS and malaria.

Despite the efforts undertaken in a difficult and fragile context and despite the considerable progress made, the country has not achieved any of the MDGs and significant challenges remain (Ministry of Planning and Follow-up of the Modernity Revolution, 2016)

The main aim of these goals is to completely eradicate poverty in the world, so that everyone can enjoy their well-being and satisfy their basic needs without difficulty. With this in mind, poverty is seen as an obstacle to development in the strict sense of the term.

It should also be noted that the concept of sustainable development raised great hopes in the 1990s, and continues to do so today. It appeared as an aid to societal reflection, a new concept capable of reorienting our actions and ultimately our society, which appears almost at the end of its tether, unable to resolve emerging issues, first and foremost the ecological question (ROBERT, 2011).

Talking about sustainable development in the DRC seems utopian, simply because its population lives on a daily basis the opposite of what the true concept recommends.

As we can see, unlike the MDGs, one of the innovations introduced in the new SDG agenda is that each country should choose, for each goal, the priority targets, taking into account its context, that will enable it to achieve said goal, and on the basis of which the country will be assessed. This approach reflects the differences between countries that have signed up to the agenda, and the firm determination of governments to contextualize each target in line with national priorities. It should also be pointed out that prioritization does not focus on the SDGs themselves, but rather on their targets.

The non-existence of a policy and the ineffectiveness of certain structures, evolving in the direction of envisaging an improvement in living conditions in the future or evolving in such a way as to prospect things according to their future in each field, never cease to discourage anyone, if it is a question of analyzing or evaluating the development situation in the DRC.

It should be noted that in 2015, the government launched the elaboration of the National Strategic Development Plan (PNSD) in order to provide itself with a medium- and long-term programmatic framework to federate the conduct of public policies and underpin economic, human and social development strategies (in line with the Sustainable Development Goals: SDGs and ¡African Union vision 2063) in the country, but due to the frequency with which the government team changed between 2016 and 2017, this development plan remains unfinalized (NSHUE,2019).

Controlling demographic growth, for example, makes it possible to identify the needs of a population in order to supply it with the products required to satisfy them. To this end, the WCED stipulates that sustainable development is only possible if demography and growth evolve in harmony with the productive potential of the ecosystem.

It is important to remember that implementing the Agenda 2030 requires unprecedented financial and technological resources. The total minimum investment required per year is estimated at $31.629 billion, or $158.14 billion for the five-year period (Ministère du Plan et Suivi de la Révolution de la Modernité, 2016).

I.5. LOCAL STRATEGY

By local strategy, it is obvious to note a typical model contributing to the resolution of the food problems of the city in particular and of the country in general. While the whole world recognizes the major possibility of feeding a large number of countries thanks to Congolese land, the DRC continues with its import strategy.

In order to achieve the sustainable development that the United Nations has called for by 2030, it's important to remember that in the DRC in general, and in each part of the country in particular, it's clear that a policy of urban and peri-urban agriculture must be applied to ensure the success of certain other objectives.

1.5.1. URBAN AND PERIURBAN AGRICULTURE
1.5.1.1. Context

According to the fifteenth session of the FAO Committee on Agriculture held in Rome from January 25-29, 1999, on urban and peri-urban agriculture, item nine on the agenda, urban agriculture, as we understand it here, refers to small areas (e.g. vacant land, gardens, orchards, balconies, various containers) used in cities to grow a few crops and raise small animals and dairy cows for personal consumption or local sales.

By peri-urban agriculture, we mean farming units close to the city that run intensive commercial or semi-commercial operations practicing horticulture (vegetables and other crops), poultry and other animal husbandry, for milk and egg production.

In the city of Kinshasa, for example, the opportunity to practice urban agriculture naturally takes two forms:

a) Soilless urban agriculture

In a densely populated city like this one, finding a free space for your garden is not at all easy, since every space has its owner. The majority of these spaces are devoted to buildings for residential, commercial or ceremonial purposes, etc. In such cases, it's obvious to consider growing on material containing mineral or organic matter to obtain the vegetables of one's choice, without counting those produced in soil, whose cultivation space and proper monitoring seem costly.

b) Urban soil-based agriculture

With this type of cultivation, any seedling or seedling used for production touches the soil directly. This type of cultivation is ideal for those who have a fairly large space in their home, or a space that is not used at all. In everyday life, many people put their trust in this type of cultivation, as it has been used for generations, particularly for commercial purposes. Thanks to this form of cultivation, the majority of people have realized their dreams.

By peri-urban agriculture, we mean farming units close to the city that run intensive commercial or semi-commercial operations practicing horticulture (vegetables and other crops), poultry and other animal husbandry, for milk and egg production.

Urban and peri-urban agriculture is practised all over the world within or around the administrative boundaries of cities. It produces products from agriculture, livestock, fisheries and forestry.

It also includes non-wood forest products, as well as the ecological functions of agriculture, fishing and forestry. Often, multiple farming and horticultural systems already exist in and around cities.

From the above, it's clear that Nouveau DAIPN has embraced the policy of urban and peri-urban agriculture, despite its lack of sophistication in adequately supplying the city's food needs.

But all the value it deserves is being overlooked by the leaders in charge, who should be supporting this strategy, but who are putting their trust in imports simply because the domain seems private to the authority of the State in place; whereas the results are improving the well-being of the Congolese people by promoting local production.

I.5.1.2. *The benefits of urban and peri-urban agriculture*

Every country in the world has an agricultural policy capable of meeting the (short- and long-term) food needs of its population without being overcome by malnutrition.

Combining all these measures is of paramount importance in the fight against food insecurity and malnutrition, but local production remains the best option in terms of contributing to food stability.

Of this local production, urban and peri-urban agriculture remains indispensable, as it remains the only good means of reassuring the availability of food in a city where it meets the need; in the supply of food of very good quality and thus contributing to the growth of the economy of the area concerned.

It's the only effective way to help reduce the rate of external food dependency and maintain good health, even in the face of climatic variations, global food price rises and recurring economic crises.

History shows that the Third World countries that most often manage to be self-sufficient in food are those that have benefited from relative geographical isolation, and where governments have had the will (and the means) to impose protectionist measures to limit imports of low-cost agricultural products from industrialized countries (India, Indonesia, Korea, etc.). Conversely, countries that have opted for free trade and specialized in exporting a few mineral products (oil) or typically tropical products (coffee, cotton, jute, groundnuts) are now largely dependent on imports for their food (Marc, 1996).

So, whether in Kinshasa, the country's other provinces, or anywhere else in the world, we still have reason to hope for an improvement in food practices that contribute to maintaining food security in urban environments, thanks to urban and peri-urban agriculture, which must undoubtedly be organized.

Indeed, professionalizing urban agriculture (market gardening), by forming ad hoc associations to which credit can be granted, may well create real wealth in Kinshasa (vegetable, fruit and tomato markets). The choice of market gardening sites must respect environmental balances (protection of fragile habitats, chemical safety through the non-excessive use of chemical pesticides and industrial fertilizers) (MUSIBONO et *al.* 2011).

Urban and peri-urban agriculture applied in a responsible urban environment will respond favorably to the needs of its population in times of crisis or other disasters that can make the whole population vulnerable; the experience of the time of containment following the COVID-19 pandemic has shown just how crucial this form of agriculture is in solving food problems for the city of Kinshasa.

In the meantime, the World Bank has had to acknowledge that African cities are not developing in the way we had imagined. Rather than growing as engines of development, it seems that many of them have become great "black holes". They do, in fact, exert an enormous power of attraction and continue to absorb rural populations at a steady pace, but it seems that the only thing that has grown in this urban space is underdevelopment itself. The city is (all too) easily summed up as a space of marginalization and exclusion, a place of shantytowns, hunger, misery and illiteracy (Filip, 2006).

CHAPTER 2: STUDY ENVIRONMENT, MATERIALS AND METHODS

11.1.STUDY ENVIRONMENT

11.1.1. Introducing the new DAIPN

The presidential agro-industrial estate of N'sele is one of the largest farms in the city of Kinshasa and in the Democratic Republic of Congo. The estate has a prestigious history in the agricultural sector in general, and in the country's agro-industrial sector in particular.

11.1.2. Geographical location

The presidential agro-industrial estate of N'sele, currently renamed NOUVEAU DAIPN, is located in the eastern part of the city-province of Kinshasa, 60 km from the city center and 30 km from N'djili international airport, on the road linking the city-province of Kinshasa and the province of KWANGO in the greater BANDUNDU region, bounded on one side by the RN1 national highway and on the other by the Congo River and the N'sele River, from which it takes its name.

11.1.3. History

The N'sele agro-industrial estate was created in 1966 as part of the struggle for economic independence for the country, which became independent on June 30, 1960. Production activities focused on agriculture, industry and livestock. In the late 1960s and early 1970s, the estate underwent considerable development, mainly as a modern presidential estate, with an imposing Chinese pagoda and the luxurious home of President Mobutu.

Around this time, a dozen or so European managers, including many Belgians and Chinese, were called in to develop model livestock and agricultural activities, including a vast battery chicken farm, cattle, pineapple and other agricultural produce. The estate included a vast animal park, Parc Président Mobutu, covering several hundred hectares, as well as enclosures housing lions, cheetahs, okapis, chimpanzees, zebras, etc., and a public Olympic swimming pool welcoming the public, including many Kinshasa residents, mainly at weekends.

To enable the estate to meet these objectives, the agricultural part of the estate was developed in 1968; however, pig rearing began in 1970, and expertise in dairy cattle rearing was acquired in 1973. Between 1966 and 1973, the estate operated as a public service.

In 1973, it was elevated to the status of a public establishment, with its own distinct heritage and industrial and commercial activities. Under this status, the estate's main tasks were to:

- ✓ Enable the country's highest authorities to organize hotel services and meetings, both national and international, at any time;
- ✓ Provide the population of Kinshasa with a varied range of basic foodstuffs (fresh milk, eggs, tinned tomatoes, pork, chicken, fish, canned fruit, etc.) that don't yet exist.

The main production unit, the main farm, covers almost 3,000 hectares, of which around 100 are used for intensive cultivation of vegetables, paddy rice, tea, manioc, tubers, corn, pineapples and various fruits.

There was also a cattle feed factory (UAB), and residential and social buildings complete the picture of this site, some of whose infrastructures were still in good condition until 2002, when periods of looting and destruction of working tools began.

11.1.4.Partnership

In 2009, the Congolese government obtained financial support from the

African Development Bank (ADB) to relaunch the farm's activities. This enabled the population of the capital Kinshasa to be served with eggs, broilers and pork, with flattering results.

The farm came to a halt due to increasing operating costs, which prevented the renewal of the herd, and the ADB funding was unable to redress the damage caused by this destruction.

In the search for a lasting solution, the Congolese state, owner of the initiative, will sign a public-private partnership with the prestigious Israeli firm "LR GROUP" on May 09, 2013, in order to rehabilitate the entire estate according to standards and relaunch all the necessary activities with a view to improving the food conditions of the Congolese population by contributing to its food security.

This time too, the results were the same and quite promising at first, but it's only after a few years that a big difference can be seen, as the stability of the domain is not assured.

11.1.5.Administration

The strategy of the new DAIPN is based on an administrative hierarchy that facilitates the coherent progress of all work. This administrative hierarchy is made up of directorates which carry out the work according to the set objectives, however, the directorates interact with each other in order to drive the field towards a common goal. As a result, the following departments are members of the Nouveau DAIPN administrative committee:

1. General Management ;
2. Secretariat ;
3. Production management ;
4. Administrative management;

5. Direction de finance ;

6. Technical direction ;

7. Sales management.

In order to better represent all these departments hierarchically or according to functional organization, it is therefore important to proceed with the representation on an organization chart to allow us to detail the divisions resulting from all these departments:

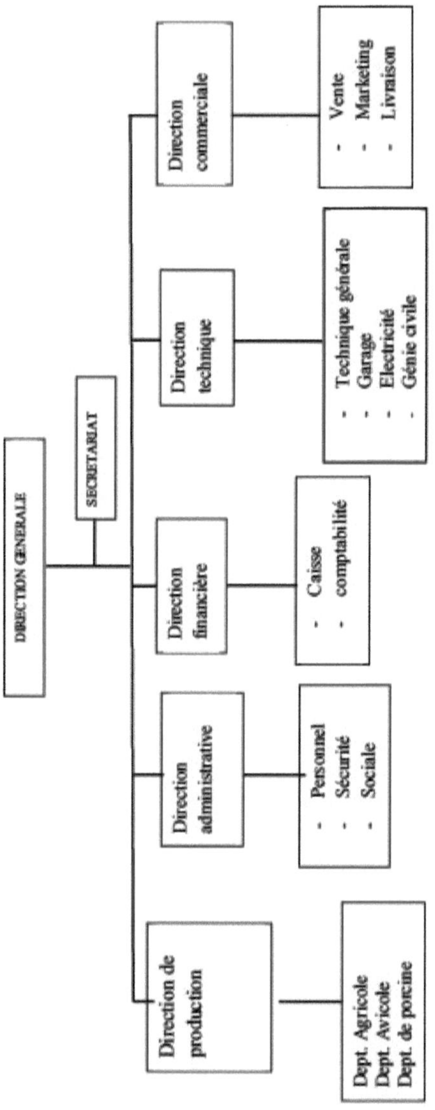

Figure 2: New DAIPN organization chart

I. 1.6. Technology

The New DAIPN remains the leading agro-industrial estate in the city and the country in general. Unlike some of the practices found in the majority of the city's agricultural areas, this one uses industrial techniques that enable large-scale production in a very short working time, thanks to the machinery and technology in place.

Thanks to this technology, the estate has no problems whatsoever with its crops, as the "LR GROUP" would have thought of exchanging its skills by installing the revolutionary drip irrigation system using water from the river and greenhouses for crops to enable regular production without being dependent on the seasons for the cultivation of tomatoes, peppers, cucumbers, eggplants, cabbages, amaranths, watermelons, melons and so on.

To ensure feed independence for its chickens, Nouveau DAIPN has set up a feed manufacturing plant that not only reduces dependence on feed, but also ensures the quality of the feed consumed by chickens and other livestock.

II. 1.7. Sales outlets

In order to sell its products effectively and reach a large number of Kinshasa's population, the Nouveau DAIPN has, in addition to direct customers such as restaurants, hotels, supermarkets,... (who are supplied according to their orders), the company has targeted sales outlets or spaces to present the results of its work and facilitate the purchase of products by retailers. Nouveau DAIPN sales outlets are located in N'SELE, KIMBANSEKE, MARCHE DE LA LIBERTE, BANDAL, SALONGO, UPN and GRAND MARCHE.

II.1.8. Slogan

As with any commercial company aiming to build customer loyalty, Nouveau DAIPN uses a phrase that reassures consumers of its products, above all by promoting and honoring the soil from which everything it produces springs.

The slogan Nouveau DAIPN uses accompanies its logo and is easily found on all its packaged products or on other objects present at the estate's points of sale. The Nouveau DAIPN slogan is "Manger frais, manger congolais" (Source: Nouveau DAIPN data).

II.2. MATERIALS AND METHODS

II.2.1. MATERIAL

Our material consists of the flagship crops of Nouveau DAIPN and all the subjects surveyed. These crops are called flagship crops because they are available on site and enable us to clearly identify their production at Nouveau DAIPN by detailing the quantity of monthly yield.

There were six (6) of these flagship crops:

1) <u>Tomato</u>
Scientific name: *Solanum lycopersicum; Family: Solanaceae*
2) <u>Peppers</u>
Scientific name: *Capsicum annuum; Family: Solanaceae*
3) <u>Cucumber</u>
Zoological Name: *Cucumis sativum; Family: Cucurbitaceae*
4) Sweet <u>eggplant</u>
Zoological Name: *Solanum melongena*; Family: *Solanaceae*
5) <u>Watermelon</u>
Zoological Name: *Citrullus lanatus; Family: Cucurbitaceae*
6) <u>Melon</u>
Scientific name: *Cucumis melo; Family: Cucurbitaceae.*

II.2.2 METHOD

The method used during our work is direct observation in the field. We also used several survey techniques such as: analysis, documentation, interview, sampling and field survey.

II.2.2.1. TECHNIQUES

Several techniques were used in our study to collect data and present them in the form of results, including:

a. Literature search

For the purposes of this study, we used the documentary research technique to gather the data and information we needed to complete this degree thesis. These data were contained in the shelves (books and scientific publications) in several of the city's libraries, and in the service reports of the new DAIPN.

b. Pre-survey

In the context of our study, the pre-investigation helped us first of all to become acquainted with the investigation environment and to identify possible problems related to the subject. It enabled us to identify preliminary aspects related to the subject, by making contact with people, some of which required in-depth study. Pre-investigation also enabled us to enrich the problem and the main hypothesis.

c. Sampling

Given the time, material and financial difficulties involved in carrying out the study, the sample size was set at 66. The subjects of the survey (consumers and retailers of the new DAIPN products) were selected at random.

d. Interview

For the purposes of our study, the interview was used to supplement the questionnaire data, by getting as many people as possible on the new DAIPN site to talk, in order to obtain additional information on the problems of food insecurity.

e. Counting and interpretation of results

Once the survey sheets had been collected, the next task was to go through each one, item by item, question by question. This work enabled us to draw up tables, as well as the figure. The results were then interpreted and discussed.

CHAPTER 3: RESULTS AND DISCUSSION

111.1.   Results

The following paragraphs present the results of this study. To help readers understand what we mean, we have divided our results into three main sections:

- Production of the new DAIPN

- General dealer information

- General consumer information

111.1.1. New DAIPN production

The New DAIPN has two main sectors:

- o The agricultural sector ;

- o The poultry or livestock sector.

III.1.1.1. ***Results by sector***

a) The agricultural sector

Table 1: Leading agricultural crops and their production

Products	November	December	January	February	Total	%
Eggplant	10349 Kg	1452 Kg	1556 Kg	5058 Kg	18415 Kg	7,4
Cucumber	14441 Kg	15079 Kg	14183 Kg	16335 Kg	60038 Kg	24,1
Peppers	15067 Kg	18857 Kg	19218 Kg	14701 Kg	67843 Kg	27,3
Tomato	55519 Kg	15955 Kg	6952 Kg	14515 Kg	92941 Kg	37,4
Watermelon	1477 Kg	-	-	5232 Kg	6709 Kg	2,7
Melon	1278 Kg	-	-	1160 Kg	2438 Kg	1,1
Grand total	**98131 Kg**	**51343**	**41909 Kg**	**57001 Kg**	**248384 Kg**	**100**

Table 1 shows that over the 4 months covered by our study, the Nouveau DAIPN produced 37.4% tomatoes (92941 Kg), 27.3% peppers (67843 Kg), 24.1% cucumbers (60038 Kg), 7.4% eggplants (18415 Kg), 2.7% watermelons (6709 Kg), and melon production accounted for only 1.1% (2438 Kg).

b) Livestock sector

The New DAIPN livestock sector is also a high-production area, but its diversity is currently characterized by fresh chickens and table eggs.

Furthermore, according to the productivity forecasts for this sector carried out at the start of the work, this area is capable of producing the following, taking into account the diversities:

Table 2: Production capacity in the livestock sector

Type of production	Daily rate	Forecasts/Tons
Broilers	3500 or 3000	1500 /day
Eggs for consumption	30 000, 60 000, 90 000	25,000,000/year
Pork meat	-	1,300/year
Fish	-	300/year

Table 2 shows that the daily rate for broiler chickens is 3,500 or 3,000 and their forecast/ton/day is 15,000; for eggs, the daily rate is 30,000, 60,000 or 90,000 and their forecast/ton/year is 25,000,000 eggs; and for pork, the forecast/ton/year is 300.

Table 3: Production in the livestock sector

Product	November	December	January	February	Total
Eggs for consumption	378000	334800	279000	125280	1117080
Percentage	34	30	25	11	100 %

As Table 3 shows, the only product that continues to exist is table eggs. In November 2019, we note the beginning of the decline with 34% productivity, 378,000 eggs or 1,050 cartons, yet the month of December, which was supposed to take over, continued to fall by up to 30%, 334,000 eggs or 930 cartons; as for January 2020, the fall becomes more and more enormous because the fall not having been controlled since the two preceding months, the remainder can only worsen, this is why we reach 25% of productivity, 279,000 eggs or 775 cartons; finally the month of February reaches the total fall of 11% with 125,280 eggs or 348 cartons.

III.1.2. GENERAL INFORMATION ON RETAILERS OF NEW DAIPN PRODUCTS

III.1.2.1. Table 4: Breakdown of subjects surveyed by profile

Gender	Frequency	Percentage
Male	6	16,7
Female	30	83,3
Total	36	100
Marital status: - Married	28	77,8
- Single	8	22,2
Total	36	100
Level of education :		
- Primary	16	44,4
- Graduate	18	50
- University study	2	5,6
Total	36	100
Age range :		
- 20 - 30 years old	12	33,3
- From 30 - 40 years	22	61,1
- From 40 - 50 years	1	2,8
- From 50 - More	1	2,8
Total	36	100

Table 4 shows that 83.3% of the subjects surveyed were female, compared with 16.7% male.

In terms of marital status, 77.8% are married and 22.2% are single. In terms of education, 50% are graduates, 44.4% have primary education and 5.6% have university education.

The age of our respondents was also taken, but by bracket. However, for our subjects surveyed, the 30 - 40 age bracket represents 61.1%; the 20 - 30 age bracket is 33.3%; the 40 - 50 age bracket 2.8% and the 50 and over age bracket 2.8% as well.

III.1.2.2. Table 5: Respondents' opinions on their initial purchases of New DAIPN products

Year of purchase	Frequency	Percentage
2010	2	5,6
2011	2	5,6
2014	6	16,7
2016	6	16,7
2017	8	22,2
2018	2	5,6
2019	4	11,1
2020	6	16,7
Total	36	100

Table 5 shows that 22.2% of respondents started buying the new DAIPN products in 2017; 16.7% in 2016; 16.7% in 2020; 16.7% in 2014; 11.1% in 2019; 5.6% in 2010; 5.6% in 2011; 5.6% in 2018.

III.1.2.3. Table 6: Respondents' opinions on their motivations for selling Nouveau DAIPN products

Motivation	Frequency	Percentage
Products appreciated by customers	*12*	*33,3*
Good quality products	*24*	*66,7*
Low purchase cost	*0*	*-*
Total	*36*	*100*

Table 6 shows that 66.7% of subjects surveyed were motivated by the good quality of New DAIPN products, and 33.3% said they were motivated by customer demand for these products. It should be noted that none of the subjects surveyed stated that they were motivated by the low cost of these products at the time of purchase.

III.1.2.4. Table 7: Survey subjects' opinions on their distance from the New DAIPN site

Distance from DAIPN site	Frequency	Percentage
Nearby retailers	16	44,4
Remote dealers	20	55,6
Total	36	100

In terms of distance from the New DAIPN site, 55.6% of resellers are remote and 44.4% are close.

III.1.2.5. Table 8: Opinions of survey subjects on their communes of origin

Municipality of origin	Frequency	Percentage
MONT-NGAFULA	2	10
Kimbanseke	8	40
GOMBE	2	10
KALAMU	2	10
KINSHASA	2	10
LIMETE	2	10
N'SELE	2	10
Total	20	100

Table 8 shows that 40% of dealers come from KIMBANSEKE, 10% from GOMBE, 10% from MONT-NGAFULA, 10% from KALAMU, 10% from KINSHASA, 10% from LIMETE and 10% from N'SELE.

III.1.2.6. Table 9: Distribution of vendors by sales outlet

Existence of a sales outlet	Frequency	Percentage
Retailers with point of sale	30	83,3
Dealers without a point of sale (operating within the delivery system)	6	17,7
Total	36	100

Table 9 shows how New DAIPN products are sold by all types of retailer. It shows that 83.3% are retailers with outlets and 17.7% do not have outlets, but use the product delivery system.

III.1.2.7. Table 10: Distribution of New DAIPN resellers by sales site

Sales sites	Frequency	Percentage
HINDOU	*12*	*40*
ZIKIDA	*12*	*40*
GOMBARE	*2*	*6,7*
N'SELE/NECROPOLE	*3*	*10*
MONT-NGAFULA	*1*	*3,3*
Total	*30*	*100*

Table 10 shows that 40% of our respondents own outlets at the HINDOU market in KIMBANSEKE; another 40% own outlets in ZIKIDA; 10% of our respondents own outlets in N'SELE near NECROPOLE; 6.7% of our respondents own outlets in the GOMBARE market near GABY bridge; and 3.3% own outlets in MONT-NGAFULA.

III.1.2.8. Table 11: Distribution of customers by product preference New DAIPN

Favorite foods	Frequency	Percentage
Peppers, tomatoes, eggplant, cucumber	36	100
Amaranth, potato leaf, cabbage, lettuce,...	0	0
Total	36	100

In terms of customer preference, retailers report that the products most in demand by their customers are peppers, tomatoes, eggplants and cucumbers. Table 11 confirms this thesis, as these products accounted for 100% of the assertion.

III.1.2.9. Customers' attitudes towards the source of products

This behavior could help consumers identify any damage that might be caused by New DAIPN products in the home, or reassure them of the quality of the products they have already chosen.

Figure 2 shows that 77.8% of the subjects surveyed confirm without doubt that their customers have no problem knowing the source of their products, and 22.2% of the subjects surveyed confirm that their customers do not know the source of their products.

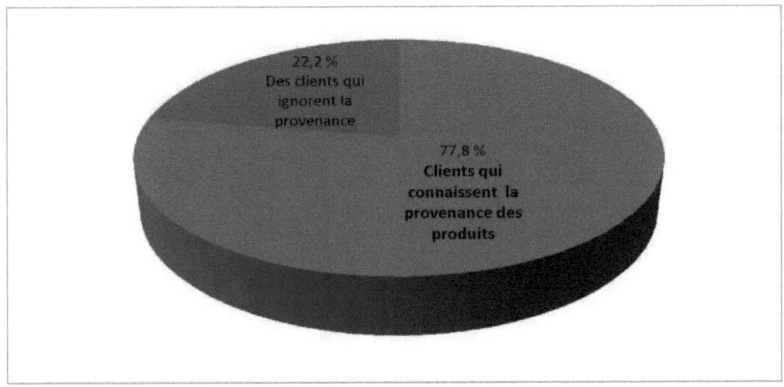

Figure 2: Customers' attitudes towards the source of their *products*

III.1.2.10. Table 12: Distribution of surveyed retailers' opinions on how they found out about the New DAIPN

Resources	Frequency	Percentage
Through friends	18	50
From himself	18	50
Through the media	0	0
Total	36	100

Table 12 shows that 50% of our respondents found out about it through their friends, and another 50% through themselves. No respondent had heard about the New DAIPN through the media.

Each of the dealers had to discover the New DAIPN products according to their efforts; this way of discovering the New DAIPN interested our investigation in order to judge the course of the marketing service.

III.1.3. GENERAL INFORMATION ON CONSUMERS OF NEW DAIPN PRODUCTS

III.1.3.1. Table 13: Breakdown of subjects surveyed by profile

Gender	Frequency	Percentage
Male	*16*	*53,3*
Female	*14*	*46,7*
Total	*30*	*100*
Marital status: - Married *- Single*	*16* *14*	*53,3* *46,7*
Total	*30*	*100*
Level of education : *- None* *- Primary*	 *4* *10*	 *13,3* *33,3*
-Graduate *-University*	*16* *-*	*44,4* *-*
Total	*30*	*100*
Age range :		
- 20 - 30 years old	*14*	*46,7*
- From 30 - 40 years	*11*	*36,6*
- From 40 - 50 years	*3*	*10*
- From 50 - More	*2*	*6,7*
Total	*30*	*100*

Table 13 shows that 53.3% of the subjects surveyed were male and 46.7% were female.

As for their marital status, 53.3% are married and 46.7% are single.

In terms of level of education, 44.4% are graduates, 33.3% have primary education and 13.3% have no education at all.

Table 13 shows that the majority of respondents are young, with 46.7% in the 20-30 age bracket, 36.6% in the 30-40 age bracket, 10% in the 40-50 age bracket and 6.7% in the 50+ age bracket.

III.1.3.2. Table 14: Opinions of surveyed consumers by year of residence in New DAIPN

Period	Frequency	Percentage
Since birth	*11*	*36,6*
1980	*3*	*10*
2000	*2*	*6,7*
2008	*2*	*6,7*
2013	*2*	*6,7*
2016	*2*	*6,7*
2019	*8*	*26,6*
Total	*30*	*100*

Table 14 shows that 36.6% of subjects surveyed have been on the New DAIPN site since birth, 26.6% since 2019, 10% since 1980, 6.7% since 2000, 6.7% since 2008, 6.7% since 2013, and 6.7% since 2016.

III.1.3.3. Table 15: Opinions of surveyed consumers on the appreciation of New DAIPN products

Quality	Frequency	Percentage
Good	*28*	*93,3*
Fairly good	*2*	*6,7*
Wrong	*0*	*0*
Total	*30*	*100*

Table 15 shows that 93.3% of consumers rated the quality of New DAIPN products as good, compared with 6.7% who rated it as fairly good. No consumer surveyed gave a contrary opinion.

III.1.3.4. Table 16: Surveyed consumers' opinions on consumption frequency

Consumption mode	*Frequency*	*Percentage*
Regular	*28*	*93,3*
By interval	*2*	*6,7*
Total	*30*	*100*

At this stage of our work, Table 16 shows 93.3% of respondents who regularly consume New DAIPN products and 6.7% of those who consume them at intervals.

III.2. DISCUSSION

Our study focused on maintaining food security for sustainable development in the DRC through a local strategy. The case of Nouveau DAIPN Kinshasa.

The results presented in the present study are discussed with those published in previous research.

In relation to the production of New DAIPN, during the months of November 2019 to February 2020, our study shows a production of 18,415 Kg or 18 tons of eggplant. This production represents only 7.4% of the total production of New DAIPN.

On the other hand, Table 1 showed a high production in the first month with 10349 Kg, the following month dropped completely, producing only 1452 Kg of eggplant. Following this horrendous drop, efforts were made to try and improve the production of this crop and the result for the third month was 1556 Kg, then 5058 Kg in the fourth month. We believe that this production instability is linked either to the rotation of cultivation areas, or to the duration of the harvest, or to pest attacks and seasonal hazards.

With regard to cucumber production, during the period of our surveys, total production was 60,038 Kg, or 24.1%. This crop yielded 14,441 Kg in the first month, then 15,079 Kg in the second, but the third month failed to maintain the balance in production and yielded 14,183 Kg, a result below that of the first month; the fourth month then contributed to the recovery of productivity with 16,335 Kg. It should therefore be recalled that this crop occupied second place in productivity, as is clearly demonstrated.

However, peppers in turn yielded a total of 67,843 Kg or 27.3% according to our survey. Table 1 clearly shows that over these four months, the first three gave encouraging productivity results. The first month produced 15067 Kg against 18857 Kg in the second month, which contributed to maintaining bell pepper productivity; the third month, therefore, produced 19218 Kg, continuing a good productive trend; but the fourth and last month fell sharply, producing 14701 Kg, lower than the first month of our study. Despite the drop in production, the quantity would still have served the population.

While tomato production during the study period was 92,941 kg, or 37.4%, and 92.5 tonnes were produced, these results encourage its cultivation and give it remarkable importance. According to the monthly production results shown in Table 1, tomatoes are one of the crops that have seen a sharp drop in production. These results show that in the first month, production was 55,519 Kg, and this remains the only one of all crops, but the drop in yield was also significant, with the second month dropping to 15,955 Kg; the third month then saw production drop to 6,952 Kg, although the results following the first month should be increasingly higher in order to better ensure the availability of this crop. The fourth month also attempted to boost tomato productivity, with a yield of 14,515 Kg.

Egg production is at least present, despite a significant drop. Total production was 1,117,080 eggs, yet the first month produced up to 378,000 eggs, or 34%, while a drop in output was felt from the second month onwards, with production at 334,800 eggs, or 30%. The third month was thus followed by the drop, with monthly production at 279,000 eggs, or 25% of this month's production, compared with 125,280 eggs in the last month, or just 11%. These declines are often linked to the age of the hens, as their renewal is often a problem at Nouveau DAIPN, which is why broilers have been emptied.

As for watermelon and melon, it's important to point out that the yields of these two crops are still supposed to be in the trial phase of their cultivation at Nouveau DAIPN, which is why production was not noted in all the months of our study, as shown in Table 1. As a result, the results of these trials still offer hope for the successful cultivation of these two fruits. These two crops share the same cropping calendar, in that they were grown for only two months, while the other two months were not productive for watermelon and melon. Table 1 shows that watermelon yielded 1477 Kg in the first month, compared with 5232 Kg in the fourth, while melon yielded 1278 Kg in the first month and 1160 Kg in the fourth. Despite this drop in production, the trials so far look promising.

Moreover, low production levels are at the root of the effects that weigh heavily on the population in the event of a food crisis at continental, global or regional level. As in some sub-Saharan African countries, the effects of the global food crisis in the DRC were felt more violently because of the low productivity of its agricultural sector and its dependence on imports of food commodities (MDG Report, 2015) cited by MUTEBA and NKULU (2019).

Note that the production of the new DAIPN is extremely low to meet food needs, even for ¼ of the population of Kinshasa. LELO NZUZI (2018) speaks of the total absence of agricultural policy in the DRC.

However, NEPAD (2013) believes that while agricultural development is insufficient to eradicate hunger and malnutrition, it is an indispensable, essential and priority element. First and foremost, increasing agricultural productivity, but also improving the efficiency of food markets, helps to reduce consumer prices, thus promoting access to food for the poorest populations, whether urban or rural. However, MUSIBONO (2006) confirms that agriculture must be the top priority in the development of a region. In his view, everything starts with agriculture.

As regards the profile of the retailers surveyed, most are married, and almost half of our sample (50%) have a humanitarian background. As MUSIBONO (2013) points out, the resellers of basic necessities are not supervised by the State via the National Beginning services, let alone by themselves via an association. As a result, this sector poses serious problems in terms of food supply. According to this author, the markets where basic necessities are sold are neither insured nor secure. BINZANGI (2013) refers to "anarchy on Kinshasa's markets".

The profiles of resellers of New DAIPN products are not regulated. Consequently, their products are not palpable in the field and/or on the market. Most of these resellers (66%) say that the New DAIPN products are of good quality, and this is their main motivation for continuing to sell them.

The majority of resellers of New DAIPN products come from downtown

Kinshasa. They cover their own travel costs from N'sele to the point of sale. However, we agree with BINZANGI (2013) that the industry needs to reach out to consumers by covering the costs of moving its products. According to this researcher, the industry must have points of sale throughout the city to ensure the proper distribution and sale of its products. MPURU (2014) talks about sales outlets for basic necessities that are officially known by customers and located in every district of the city. However, products from the New DAIPN are virtually absent, unknown and, above all, confused with other products on Kinshasa's markets.

With regard to the profiles of consumers of New DAIPN products. Most of the subjects surveyed have a medium or humanitarian level of education, and there are no university graduates.

BINZANGI (2013) talks about the dietary ignorance of Congolese intellectuals. People who have reached a high level of education or university don't seem to know or understand the importance of consuming organic elements in their bodies. They consume fresh (non-bio) food, unaware of all the consequences this can have on their bodies in the short, medium and long term.

Almost all (93.3%) of these consumers confirm that products from the New DAIPN are of very good quality, and they consume them regularly.

Let's consider with MUSIBONO (2012) that the more people consume good quality organic food, the greater their longevity and even their life expectancy. The Congolese population in general, and Kinshasa in particular, is a slave to imported fresh food that is very poorly preserved. As a result, Congolese urban populations in particular are very fragile and vulnerable to the various threats of disease and/or epidemics, thus increasing the mortality rate.

The food problems in Kinshasa are due to the culture adopted by

Kinshasa's self-proclaimed city-dwellers. The people of Kinshasa prefer to eat food from unknown places, of dubious quality and preservation. As mentioned above, this phenomenon exposes the population to a number of threats.

Conclusion

Eliminating food insecurity remains the overriding wish of any state that aims to ensure the well-being of its population. But achieving this goal is becoming increasingly complicated for most of the world's countries, especially those in the developing world, including the Democratic Republic of Congo.

Food insecurity is a scourge that affects the whole of the Democratic Republic of the Congo, and Kinshasa in particular is just as hard hit. Yet the country has everything it takes to tackle this scourge, with its abundance of animal and plant species that are useful for consumption but remain unexploited in sufficiently large quantities to meet demand.

The fight against hunger is a major preoccupation for many countries in the fight against poverty, and is also the focus of several international organizations.

To better tackle this obstacle of food insecurity, it is therefore up to the State to make a firm commitment in its policy, which should guide the realization of all the objectives that will be agreed thanks to the close collaboration that it will have to maintain with the scientists who will set the paths to follow.

Success in feeding a nation is a major step towards sustainable development, as many important aspects are linked to its success. As a result, there is no doubt that most projects struggle to succeed due to certain aspects that are omitted from the outset, including the extreme poverty of the majority of the population, who are increasingly weakened by the lack of appropriate food for their development.

Indeed, the population of Kinshasa still lives in food insufficiency, despite the local production of foodstuffs in Kinshasa by small breeders, market gardeners, the New DAIPN, etc. This is still inferior to the needs of each individual and also to the availability, access, use and stability of all the products produced by them. The population increasingly wants to consume their products, but unfortunately the lack of organization in the distribution of all goods means that consumer demand cannot be met; as a result, all social strata face enormous challenges in benefiting from or regularly consuming food considered to be of good quality and locally produced.

Furthermore, the Congo, our beautiful country with all its potential, will have to correct its food strategy in the not-too-distant future, in order to boost its development and guarantee a better future for future generations. This strategy will help it to position itself well for sustainable development and to maintain the social well-being of its entire population, whose hope for a new life has remained unfulfilled for decades.

Maintaining food security requires, above all, a political program to support this strategy by boosting productivity at both local and national levels throughout the country. New structures must be created through this agricultural policy, with the aim of creating new jobs and contributing to the availability of food.

Initially, the New DAIPN produced a wide variety of vegetables and fruit, making it a major supplier of food products to the city, but a few years later, productivity declined and some crops disappeared due to a lack of resources on this large estate.

At this stage, Nouveau DAIPN alone will not be able to cover the food needs of the whole city of Kinshasa, but it can remain one of the major suppliers of some essential food products, despite its low productivity compared to the supply of the whole city.

In this work, we have shown the monthly productivity of some crops produced at the New DAIPN, and despite the effort that the latter makes to feed the population, improvements in yield are necessary in order to transform all the monthly production present into daily production, because up to this level, all production remains insignificant to meet the needs of a population estimated at 12 million inhabitants (The case of Kinshasa).

Finally, the policy of food security remains the only sure way for the Democratic Republic of Congo to ensure sustainable development, but this will require the promotion of large-scale local production.

In addition to its great mining potential, our country has enormous potential in agriculture, livestock farming, fishing, tourism, etc., all of which need to be well organized if we are to succeed in achieving the MDG2.

RECOMMENDATIONS

The researcher's proposals are essential for the completion of his study, and above all for his target audience, who can only wait for new measures to improve his work.

As a presidential domain under the direction of the State, Nouveau DAIPN must function correctly in order to improve the social situation so deplored by the Congolese people in general and the people of Kinshasa in particular, with a view to achieving sustainable food security.

The fact that the DRC is classified as one of the world's most food-insecure countries is a serious lack of consideration for this great people; it only proves the disorganization at the top of the country's agricultural sector, especially as it is recognized as having great potential in this area for the continent and the world. The same disorganization has led Nouveau DAIPN to find itself sometimes unable to ensure the proper distribution of its products to every corner of the city.

For this work, we suggest the following:

1. TO THE STATE:
    - Organize the agricultural sector by allocating a responsible budget;
    - Support local production to minimize food imports;
    - Maintain agricultural and/or rural roads so that essential products can be quickly evacuated to urban areas in high demand;
    - Minimize certain taxes so that sales on the market take place with due regard for the purchasing power of the already poor population;
    - To ensure the quality, hygiene and preservation of products from the point of production to the point of sale;
    - Encourage scientific research, especially by young people, in the agricultural sector via technical agricultural institutions.

2. TO THE NEW DAIPN:

- Increase production by diversifying crops;

- Use qualified manpower and acceptable technical resources;

- Know your customers' needs on a regular basis;

- Secure the market by setting up its own sales outlets;

- Increase the number of sales outlets;

- Transport products from the production site to the various points of sale;

- Establish an effective marketing department to raise awareness of the importance of organic consumption;

- Set the cost of products on the market, taking into account the population's purchasing power;

- Identify all resellers of New DAIPN products;

- Create partnerships with small breeders and farmers;

3. TO THE PUBLIC

- Eat organic and local food;

- Avoid consumption of fresh food;

- Ensure good hygiene and conservation of food products;

- Trust local consumers; etc.

BIBLIOGRAPHY

ACF-IN (2008): Introduction à la sécurité alimentaire - principes d'intervention, ACF-IN, Paris.

ALLEN H. (1998): What world for tomorrow? Scenarios for the 21st^e century, translated from the American by Monique Berry, NOUVEAUX HORIZONS.

BINZANGI L. (2014) : " Réflexions sur l'évolution de l'environnement de Kinshasa : d'une portion biosphérique à une " cupidosphère ", In Cahiers Congolais de ¡'Aménagement et du Bâtiment (N°003), IBTP. Kinshasa, pp. 83-91.

BUGEME D., ULIMWENGU J. (2019): Climate-smart agriculture and food security in the Democratic Republic of Congo, journal Congo challenge, Volume 1 Issue 1, July 2019.

WCED (1987): Our Common Future: The Brundtland Report.

DUMBI C. et al. (2016): What future for market gardening households? Conjonctures congolaises, Kinshasa.

FAO (1996): The State of Food and Agriculture - Macroeconomic Dimensions of Food Security, FAO, Rome, Italy.

FAO (2010): The state of food insecurity in the world combating food insecurity in protracted crises, FAO, Rome, Italy.

FAO, FIDA, WHO, WFP & UNICEF (2018): The state of food security and nutrition in the world strengthening climate change resilience for food security and nutrition. FAO, Rome, Italy.

FILIP D., Jean-Pierre J. (2006): La ville de Kinshasa, une architecture du verbe. Editions Esprit, Paris.

FISCRCR (2005): How to assess food security? A practical guide for African national societies. Geneva, Switzerland.

GAUDREAULT V. (2011): Analysis of urban agriculture in large urban centers in North America. Essay presented at the Centre Universitaire de Formation en Environnement de l'Université de Sherbrooke for the Master's degree in Environment (M. Env.).

GOOSSENS F. (1997): Aliments dans les villes - Rôle des SADA dans la sécurité alimentaire de Kinshasa. FAO, Rome, Italy.

HUART A., NKIDIATA O. (2004): La situation de l'élevage de volaille en RDC et à Kinshasa. Eco Congo, Kinshasa.

LELO F. (2011), Kinshasa-Planification et aménagement, ed. L'Harmattan, Paris.

MALELE S. (2003): Situation des ressources génétiques forestières de la RDC, working paper FGR/56, FAO, Rome, Italy.

MARC D. (1996): Les projets de développement agricole-Manuel d'expertise, éditions KARTHALA, Paris, France.

MARC D. (2004): Agriculture et paysanneries des Tiers mondes, éditions KARTHALA, Paris, France.

MPUPU (2012) cited in MPUPU et al (2019): Local chicken (Gallus gallus domesticus L.) farming in the Democratic Republic of Congo: issues on food security and climate change. Revue Africaine d'Environnement et d'Agriculture 2019; 2(1), 76-83

MUKOKA F. (2014): Discours et pratiques du développement au Congo: interrogation et ré interrogation politologique. Editions MES, Kinshasa.

MUSIBONO D. (2009), la RDC face aux enjeux de la géostratégie des ressources naturelles, ed. L'Harmattan, Paris.

MUSIBONO D., BIEY E., KISANGALA M., NSIMANDA C., MUNZUNDU B., KEKOLEMBA V., & PAULUS J. (2011): Urban agriculture as a response to unemployment in Kinshasa, Democratic Republic of Congo, [VertigO] La revue électronique en sciences de l'environnement, Vertigo, Volume 11, Numéro 1, May 2011.

MUSIBONO D. (2012), Politiques toxiques et pollution de l'environnement mondial. A collective suicide, ed. ERGS-Kinshasa.

MUSIBONO D. (2015), Durabilité socio-environnementale de l'industrie extractive en Afrique: le paradoxe congolais, ed. ERGS-Kinshasa.

MUSIBONO D. (2016): Informations environnementales, course en premier Graduat des Sciences de l'environnement, Facultés de Sciences de l'Université de Kinshasa.

MUTEBA D., NKULU J. (2019): Food crises and mitigation measures in the Democratic Republic of Congo. Review of strategies and promotion of good practices. MEDIAS PAUL, Kinshasa.

NDONGO M. (2017): Analysis and implementation of efficient mesological education as a guarantee of sustainable development in the DRC.
A few schools in the LIVULU district of Kinshasa. Final thesis in
Environmental Sciences at the University of Kinshasa.

NEPAD (2013): African agricultures, transformations and perspectives, NEPAD, November 2013.

NSHUE A. (2019): Regards sur l'économie congolaise de 2015 à 2018, Congo challenge magazine, Volume 1 Issue 1, July 2019.

WFP (2014): DRC In-depth food security and vulnerability analysis. WFP, Rome, Italy.

RDC, MINISTERE DU PLAN ET SUIVI DE LA REVOLUTION DE LA MODERNITE (2016): Contextualization and Prioritization of the Sustainable Development Goals (SDGs) in the Democratic Republic of Congo.

RIVERA F., RIVERA M. (1991): Planification agricole alimentaire en fonction des besoins nutritionnels de la population: surface nécessaire pour la production des aliments, KARTHALA-ACCT-AUPELF, Paris.

ROBERT J. (2011): Durables?, éditions Recherches, Paris.

TOUZARD, J.-M. & FOURNIER, S. (2014). The complexity of food systems: an asset for food security? [VertigO] La revue électronique en sciences de l'environnement, Volume 14 Number 1 | May 2014.

APPENDICES

Image 1: Collection of eggs for consumption at Nouveau DAIPN

Image 2: Tomato produced at Nouveau
DAIPN

Image 3: New DAIPN eggplant

Image 4: New DAIPN green tomato

Image 5: New DAIPN cucumber

Image 6: New DAIPN bell pepper

Image 7: Some key crops in the

I.   PROFIL DE L'ENQUETE (CONSOMMATEURS)

1. Sexe : M [   ]   F [   ]
2. Tranche d'âge :
    - 20 – 30 ans [   ]
    - 30 – 40 ans [   ]
    - 40 – 50 ans [   ]
    - 50 – plus [   ]
3. Etat civil :
    - Célibataire [   ]
    - Marié (e) [   ]
    - Divorcé (e) [   ]
    - Veuf (ve) [   ]
4. Niveau d'instruction :
    - Néant [   ]
    - Primaire [   ]
    - Secondaire [   ]
    - Diplômé [   ]
    - Supérieur et universitaire [   ]
    - Post universitaire [   ]
5. Valeur des revenus mensuels :
    - 1 – 50 $ [   ]
    - 51 – 100$ [   ]
    - 101 – 200$ [   ]
    - 201 – 300$ [   ]
    - 301 – 400$ [   ]
    - 401 – 500$ [   ]
    - 500 à plus [   ]

II.   QUESTIONS

1) Depuis quelle année habitez-vous la cité DAIPN ?

R)...

2) Connaissez – vous les produits issus de Nouveau DAIPN ?

Oui [   ]   Non [   ]

-   Si Oui, sont-ils de quelle qualité ?

Bonne [____]   Assez bonne [____]   Mauvaise [____]

- Si Oui, sont-ils vendus à un coût acceptable ?

Oui [____]   Non [____]

3) Consommez-vous les produits de DAIPN ?

Oui [____]   Non [____]

Si Oui, quels produits consommez-vous régulièrement ?

R)..

Et pourquoi...

Si Non, pourquoi ?..

Si Non, quelle est la source de provenance des produits que vous consommez ?

....

4) Avez-vous des suggestions à faire à DAIPN ?

Oui [____]   Non [____]

Si Oui, lesquelles ?

....

....

....

....

....

Merci

QUESTIONNAIRE D'ENQUETE SUR LE TRAVAIL DE FIN D'ETUDE INTUTILE
« CONTRIBUTION A L'ETUDE SUR LE MAINTIEN DE LA SECURITE ALIMENTAIRE EN RDC PAR LA
STRATEGIE LOCALE. CAS DE NOUVEAU DAIPN A KINSHASA ».
Par l'étudiant NDONGO NDJONDJO Michel de deuxième licence en Sciences de
l'Environnement /UNIKIN

I.   PROFIL DE L'ENQUETE (REVENDEURS)

6. Sexe : M ☐   F ☐

7. Tranche d'âge :
- 20 – 30 ans ☐
- 30 – 40 ans ☐
- 40 – 50 ans ☐
- 50 – plus ☐

8. Etat civil :
- Célibataire ☐
- Marié (e) ☐
- Divorcé (e) ☐
- Veuf (ve) ☐

9. Niveau d'instruction :
- Néant ☐
- Primaire ☐
- Secondaire ☐
- Diplômé ☐
- Supérieur et universitaire ☐
- Post universitaire ☐

10. Valeur des revenus mensuels :
- 1 – 50 $ ☐
- 51 – 100$ ☐
- 101 – 200$ ☐
- 201 – 300$ ☐
- 301 – 400$ ☐
- 401 – 500$ ☐
- 500 à plus ☐

II.   QUESTIONS

1) Depuis quelle période achetez-vous les produits DAIPN ?

De 1970 – 1990 ☐   1991 - 2000 ☐   2001 – 2010 ☐   2011 – 2020 ☐

Autres, à préciser...

2) Quelles sont les motivations à vendre ces produits ?

a. Produits trop sollicités par les clients [ ] b. produits de bonne qualité [ ] c. faible coût à l'achat [ ] d. A préciser...

3. Habitez-vous proche d'un point de vente des produits DAIPN ?

Oui [ ] Non [ ]

Si Non, habitez-vous quelle commune de Kinshasa ?...

4. Avez-vous un point de vente connu pour vos produits à la cité ?

Oui [ ] Non [ ]

Si Oui, où se trouve-t-il ?..

Si Non, comment vendez- vous vos produits ?..

5. Quels sont les produits les plus demandés par vos clients ?

R)..

....

Et
pourquoi ?..

....

6. Vos clients, connaissent-ils la provenance des produits qu'ils consomment ?

Oui [ ] Non [ ]

7. Comment avez-vous connu DAIPN ?...

8. Avez-vous des suggestions à faire à DAIPN ?

Oui [ ] Non [ ]

Si Oui, lesquelles ?

....

....

....

....

....

....

Merci

I want morebooks!

Buy your books fast and straightforward online - at one of world's fastest growing online book stores! Environmentally sound due to Print-on-Demand technologies.

Buy your books online at
www.morebooks.shop

Kaufen Sie Ihre Bücher schnell und unkompliziert online – auf einer der am schnellsten wachsenden Buchhandelsplattformen weltweit! Dank Print-On-Demand umwelt- und ressourcenschonend produziert.

Bücher schneller online kaufen
www.morebooks.shop

Printed by Books on Demand GmbH, Norderstedt / Germany